Martin Luther King Jr.
A Great Leader

Columbus, OH

SRAonline.com

Send all inquiries to this address:
SRA/McGraw-Hill
4400 Easton Commons
Columbus, OH 43219

ISBN: 978-0-07-606661-2
MHID: 0-07-606661-4

1 2 3 4 5 6 7 8 9 NOR 13 12 11 10 09 08 07

Martin Luther King Jr. was a great man.
He improved the lives of African Americans.
He cared about the poor. He spent his life
helping people gain their rights.

King was born in 1929 in Atlanta, Georgia. Life in the South was not the same as it is today. The laws then said that blacks and whites could not mix.

Black and white people could not go to the same schools. They could not stay in the same hotels. They had to eat separately in restaurants. They could not even drink from the same water fountains.

King grew up in a close family. His mom
and dad taught him that all people should
have the same rights. He learned to respect
people whether they were rich or poor.

King went to college. Then he married Coretta Scott. They lived in Montgomery, Alabama. King was a minister. Both his father and grandfather were ministers too.

In 1955 a black woman named Rosa
Parks had a problem. She would not give up
her seat on a bus to a white man. This was
against the law. She went to jail.

African Americans were angry. King led them in a plan to stay off the buses. They would not ride until they could sit in any seat they chose.

King and his family got threats. He was arrested. He never changed his mind, though. After 382 days black people won the right to sit where they wanted. They felt great excitement!

Black people kept up the fight for their rights. African American students wanted to be served at lunch counters. In 1960 King joined them in Atlanta.

King was arrested again. This time he went to jail. He usually stayed calm. While some people did suggest that violence would help their cause, King did not agree.

In 1963 King led a peaceful march in
Washington, D.C. His speech sent a thrill
through the huge crowd.

The next year King won the Nobel
Peace Prize.

In 1968 King went to Memphis,
Tennessee, to help workers on strike.
While he was there he was shot and killed.
We have not forgotten his vital work for
people's rights, though.

Today we honor Martin Luther King Jr.
in many ways. The nation celebrates his
birthday in January. Many of our cities have
schools and streets named for him.

Vocabulary

problem (prob´ ləm) (page 8) *n*. A personal difficulty; a tricky or uncomfortable situation.

changed mind (chānjd mīnd) (page 10) *v*. Went back on a decision.

excitement (ik sīt´ mənt) (page 10) *n*. A mood or feeling of high interest or energy; delight; joy.

usually (ū´ zhōō əl ē) (page 12) *adv*. Most of the time.

suggest (səg jest´) (page 12) *v*. To give or tell an idea.

thrill (thril) (page 13) *n*. A feeling of excitement.

Comprehension Focus: Asking Questions

1. What questions did you ask yourself as you began to read the book?
2. What answers did you find?